Engineering and Construction Contract

This contract should be used for the appointment of a contractor for engineering and construction work, including any level of design responsibility

Option E: Cost reimbursable contract

An NEC document

June 2005
(with amendments June 2006)

OGC endorsement of NEC3

OGC advises public sector procurers that the form of contract used has to be selected according to the objectives of the project, aiming to satisfy the *Achieving Excellence in Construction* (AEC) principles.

This edition of the NEC (NEC3) complies fully with the AEC principles. OGC recommends the use of NEC3 by public sector construction procurers on their construction projects.

NEC is a division of Thomas Telford Ltd, which is a wholly owned subsidiary of the Institution of Civil Engineers (ICE), the owner and developer of the NEC.

The NEC is a family of standard contracts, each of which has these characteristics:

- Its use stimulates good management of the relationship between the two parties to the contract and, hence, of the work included in the contract.
- It can be used in a wide variety of commercial situations, for a wide variety of types of work and in any location.
- It is a clear and simple document – using language and a structure which are straightforward and easily understood.

NEC3 Engineering and Construction Contract is one of the NEC family and is consistent with all other NEC3 documents. Also available are the Engineering and Construction Contract Guidance Notes, Flow Charts and Options A, B, C, D, E and F.

ISBN (complete box set) 978 07277 3382 5
ISBN (this document) 978 07277 3364 1
ISBN (Engineering and Construction Contract) 978 07277 3359 7
ISBN (Engineering and Construction Contract Guidance Notes) 978 07277 3366 5
ISBN (Engineering and Construction Contract Flow Charts) 978 07277 3367 2
ISBN (Option A: Priced contract with activity schedule) 978 07277 3360 3
ISBN (Option B: Priced contract with bill of quantities) 978 07277 3361 0
ISBN (Option C: Target contract with activity schedule) 978 07277 3362 7
ISBN (Option D: Target contract with bill of quantities) 978 07277 3363 4
ISBN (Option F: Management contract) 978 07277 3365 8

Consultative edition 1991
First edition 1993
Second edition November 1995
Reprinted with corrections May 1998
Third edition June 2005
Reprinted with amendments June 2006, 2007

Cover photo, Golden Jubilee Bridge, courtesy of City of Westminster

9 8 7 6 5 4 3

British Library Cataloguing in Publication Data for this publication is available from the British Library.

Typeset by Academic + Technical, Bristol

Printed and bound in Great Britain by Bell & Bain Limited, Glasgow, UK

CONTENTS

In this contract the core clauses are the ECC core clauses and the clauses set out in the ECC as main Option clauses; Option E. The latter are included in sequence and are printed in **bold type** in this contract.

ACKNOWLEDGEMENTS

The NEC first edition was produced by the Institution of Civil Engineers through its NEC Working Group.

The original NEC was designed and drafted by Dr Martin Barnes then of Coopers and Lybrand with the assistance of Professor J. G. Perry then of the University of Birmingham, T. W. Weddell then of Travers Morgan Management, T. H. Nicholson, Consultant to the Institution of Civil Engineers, A. Norman then of the University of Manchester Institute of Science and Technology and P. A. Baird, then Corporate Contracts Consultant, Eskom, South Africa.

The second edition of the NEC documents for engineering and construction contracts was produced by the Institution of Civil Engineers through its NEC Panel.

The third edition of the NEC Engineering and Construction Contract was produced by the Institution of Civil Engineers through its NEC Panel. The Flow Charts were produced by John S. Gillespie with assistance from Tom Nicholson.

The members of the NEC Panel are:

P. Higgins, BSc, CEng, FICE, FCIArb (Chairman)
P. A. Baird, BSc, CEng, FICE, M(SA)ICE, MAPM
M. Barnes, BSc(Eng), PhD, FREng, FICE, FCIOB, CCMI, ACIArb, MBCS, FInstCES, FAPM
A. J. Bates, FRICS, MInstCES
A. J. M. Blackler, BA, LLB(Cantab), MCIArb
P. T. Cousins, BEng(Tech), DipArb, CEng, MICE, MCIArb, MCMI
L. T. Eames, BSc, FRICS, FCIOB
F. Forward, BA(Hons), DipArch, MSc(Const Law), RIBA, FCIArb
Professor J. G. Perry, MEng, PhD, CEng, FICE, MAPM
N. C. Shaw, FCIPS, CEng, MIMechE
T. W. Weddell, BSc, CEng, DIC, FICE, FIStructE, ACIArb

NEC Consultant:

R. A. Gerrard, BSc(Hons), MRICS, FCIArb, FInstCES

Secretariat:

A. Cole, LLB, LLM, BL
J. M. Hawkins, BA(Hons), MSc
F. N. Vernon (Technical Adviser), BSc, CEng, MICE

The Institution of Civil Engineers acknowledges the help in preparing the third edition given by many other people, in particular, by:

J. C. Broome, BEng
A. Else, BSc, CEng, FICE
C. Flook, Esq.
R. Lewendon, FICE, MIHT, MCIArb, MAPM
T. H. Nicholson, BSc, FICE
R. Patterson, BA, MA(Cantab), MBA, CEng, MICE
C. Reed, CEng, MA, MSc, FICE
D. Weeks, FRICS
S. Zarka, FRICS

SCHEDULE OF OPTIONS

	One of the following dispute resolution Options must be selected to complete the chosen main Option.
Option W1	Dispute resolution procedure (used unless the United Kingdom Housing Grants, Construction and Regeneration Act 1996 applies).
Option W2	Dispute resolution procedure (used in the United Kingdom when the Housing Grants, Construction and Regeneration Act 1996 applies).
	The following secondary Options should then be considered. It is not necessary to use any of them. Any combination other than those stated may be used.
Option X2	Changes in the law
Option X4	Parent company guarantee
Option X5	Sectional Completion
Option X6	Bonus for early Completion
Option X7	Delay damages
Option X12	Partnering
Option X13	Performance bond
Option X14	Advanced payment to the *Contractor*
Option X15	Limitation of the *Contractor*'s liability for his design to reasonable skill and care
Option X16	Retention
Option X17	Low performance damages
Option X18	Limitation of liability
Option X20	Key Performance Indicators (not used with Option X12)
	The following Options dealing with national legislation should be included if required.
Option Y(UK)2	The Housing Grants, Construction and Regeneration Act 1996
Option Y(UK)3	The Contracts (Rights of Third Parties) Act 1999
Option Z	*Additional conditions of contract*
Note	Options X1, X3, X8 to X11, X19 and Y(UK)1 are not used.

AMENDMENTS JUNE 2006

The following amendments have been made to the June 2005 edition.

Page	Clause	Line	
11	32.1	4	deleted: 'and of notified early warning matters'
45	Contract Data Part one, 1.	3	added, '(with amendments June 2006)'
48	Contract Data Part one, 1.	23	deleted: 'If there are additional compensation events' and following 4 lines

CORE CLAUSES

1 General

Actions 10

10.1 The *Employer*, the *Contractor*, the *Project Manager* and the *Supervisor* shall act as stated in this contract and in a spirit of mutual trust and co-operation.

Identified and defined terms 11

11.1 In these conditions of contract, terms identified in the Contract Data are in italics and defined terms have capital initials.

11.2 (1) The Accepted Programme is the programme identified in the Contract Data or is the latest programme accepted by the *Project Manager*. The latest programme accepted by the *Project Manager* supersedes previous Accepted Programmes.

(2) Completion is when the *Contractor* has

- done all the work which the Works Information states he is to do by the Completion Date and
- corrected notified Defects which would have prevented the *Employer* from using the *works* and Others from doing their work.

If the work which the *Contractor* is to do by the Completion Date is not stated in the Works Information, Completion is when the *Contractor* has done all the work necessary for the *Employer* to use the *works* and for Others to do their work.

(3) The Completion Date is the *completion date* unless later changed in accordance with this contract.

(4) The Contract Date is the date when this contract came into existence.

(5) A Defect is

- a part of the *works* which is not in accordance with the Works Information or
- a part of the *works* designed by the *Contractor* which is not in accordance with the applicable law or the *Contractor*'s design which the *Project Manager* has accepted.

(6) The Defects Certificate is either a list of Defects that the *Supervisor* has notified before the *defects date* which the *Contractor* has not corrected or, if there are no such Defects, a statement that there are none.

(7) Equipment is items provided by the *Contractor* and used by him to Provide the Works and which the Works Information does not require him to include in the *works*.

(8) The Fee is the sum of the amounts calculated by applying the *subcontracted fee percentage* to the Defined Cost of subcontracted work and the *direct fee percentage* to the Defined Cost of other work.

(9) A Key Date is the date by which work is to meet the Condition stated. The Key Date is the *key date* stated in the Contract Data and the Condition is the *condition* stated in the Contract Data unless later changed in accordance with this contract.

(10) Others are people or organisations who are not the *Employer*, the *Project Manager*, the *Supervisor*, the *Adjudicator*, the *Contractor* or any employee, Subcontractor or supplier of the *Contractor*.

(11) The Parties are the *Employer* and the *Contractor*.

(12) Plant and Materials are items intended to be included in the *works*.

(13) To Provide the Works means to do the work necessary to complete the *works* in accordance with this contract and all incidental work, services and actions which this contract requires.

(14) The Risk Register is a register of the risks which are listed in the Contract Data and the risks which the *Project Manager* or the *Contractor* has notified as an early warning matter. It includes a description of the risk and a description of the actions which are to be taken to avoid or reduce the risk.

(15) The Site is the area within the *boundaries of the site* and the volumes above and below it which are affected by work included in this contract.

(16) Site Information is information which

- describes the Site and its surroundings and
- is in the documents which the Contract Data states it is in.

(17) A Subcontractor is a person or organisation who has a contract with the *Contractor* to

- construct or install part of the *works*,
- provide a service necessary to Provide the Works or
- supply Plant and Materials which the person or organisation has wholly or partly designed specifically for the *works*.

(18) The Working Areas are those parts of the *working areas* which are

- necessary for Providing the Works and
- used only for work in this contract

unless later changed in accordance with this contract.

(19) Works Information is information which either

- specifies and describes the *works* or
- states any constraints on how the *Contractor* Provides the Works

and is either

- in the documents which the Contract Data states it is in or
- in an instruction given in accordance with this contract.

(23) Defined Cost is

- **the amount of payments due to Subcontractors for work which is subcontracted without taking account of amounts deducted for**
  - **retention,**
  - **payment to the *Employer* as a result of the Subcontractor failing to meet a Key Date,**
  - **the correction of Defects after Completion,**
  - **payments to Others and**
  - **the supply of equipment, supplies and services included in the charge for overhead cost within the Working Areas in this contract**

and

- **the cost of components in the Schedule of Cost Components for other work**

less Disallowed Cost.

(25) Disallowed Cost is cost which the *Project Manager* decides

- **is not justified by the *Contractor*'s accounts and records,**
- **should not have been paid to a Subcontractor or supplier in accordance with his contract,**
- **was incurred only because the *Contractor* did not**
  - **follow an acceptance or procurement procedure stated in the Works Information or**
  - **give an early warning which this contract required him to give**

and the cost of

- **correcting Defects after Completion,**
- **correcting Defects caused by the *Contractor* not complying with a constraint on how he is to Provide the Works stated in the Works Information,**
- **Plant and Materials not used to Provide the Works (after allowing for reasonable wastage) unless resulting from a change to the Works Information,**
- **resources not used to Provide the Works (after allowing for reasonable availability and utilisation) or not taken away from the Working Areas when the *Project Manager* requested and**
- **preparation for and conduct of an adjudication or proceedings of the *tribunal*.**

(29) The Price for Work Done to Date is the total Defined Cost which the *Project Manager* forecasts will have been paid by the *Contractor* before the next assessment date plus the Fee.

(32) The Prices are the Defined Cost plus the Fee.

Interpretation and the law 12

12.1 In this contract, except where the context shows otherwise, words in the singular also mean in the plural and the other way round and words in the masculine also mean in the feminine and neuter.

12.2 This contract is governed by the *law of the contract*.

12.3 No change to this contract, unless provided for by the *conditions of contract*, has effect unless it has been agreed, confirmed in writing and signed by the Parties.

12.4 This contract is the entire agreement between the Parties.

Communications 13

13.1 Each instruction, certificate, submission, proposal, record, acceptance, notification, reply and other communication which this contract requires is communicated in a form which can be read, copied and recorded. Writing is in the *language of this contract*.

13.2 A communication has effect when it is received at the last address notified by the recipient for receiving communications or, if none is notified, at the address of the recipient stated in the Contract Data.

13.3 If this contract requires the *Project Manager*, the *Supervisor* or the *Contractor* to reply to a communication, unless otherwise stated in this contract, he replies within the *period for reply*.

13.4 The *Project Manager* replies to a communication submitted or resubmitted to him by the *Contractor* for acceptance. If his reply is not acceptance, the *Project Manager* states his reasons and the *Contractor* resubmits the communication within the *period for reply* taking account of these reasons. A reason for withholding acceptance is that more information is needed in order to assess the *Contractor*'s submission fully.

13.5 The *Project Manager* may extend the *period for reply* to a communication if the *Project Manager* and the *Contractor* agree to the extension before the reply is due. The *Project Manager* notifies the *Contractor* of the extension which has been agreed.

13.6 The *Project Manager* issues his certificates to the *Employer* and the *Contractor*. The *Supervisor* issues his certificates to the *Project Manager* and the *Contractor*.

13.7 A notification which this contract requires is communicated separately from other communications.

13.8 The *Project Manager* may withhold acceptance of a submission by the *Contractor*. Withholding acceptance for a reason stated in this contract is not a compensation event.

The *Project Manager* and the *Supervisor* 14

14.1 The *Project Manager*'s or the *Supervisor*'s acceptance of a communication from the *Contractor* or of his work does not change the *Contractor*'s responsibility to Provide the Works or his liability for his design.

14.2 The *Project Manager* and the *Supervisor*, after notifying the *Contractor*, may delegate any of their actions and may cancel any delegation. A reference to an action of the *Project Manager* or the *Supervisor* in this contract includes an action by his delegate.

14.3 The *Project Manager* may give an instruction to the *Contractor* which changes the Works Information or a Key Date.

14.4 The *Employer* may replace the *Project Manager* or the *Supervisor* after he has notified the *Contractor* of the name of the replacement.

Adding to the Working Areas 15

15.1 The *Contractor* may submit a proposal for adding an area to the Working Areas to the *Project Manager* for acceptance. A reason for not accepting is that the proposed area is either not necessary for Providing the Works or used for work not in this contract.

Early warning 16

16.1 The *Contractor* and the *Project Manager* give an early warning by notifying the other as soon as either becomes aware of any matter which could

- increase the total of the Prices,
- delay Completion,
- delay meeting a Key Date or
- impair the performance of the *works* in use.

The *Contractor* may give an early warning by notifying the *Project Manager* of any other matter which could increase his total cost. The *Project Manager* enters early warning matters in the Risk Register. Early warning of a matter for which a compensation event has previously been notified is not required.

16.2 Either the *Project Manager* or the *Contractor* may instruct the other to attend a risk reduction meeting. Each may instruct other people to attend if the other agrees.

16.3 At a risk reduction meeting, those who attend co-operate in

- making and considering proposals for how the effect of the registered risks can be avoided or reduced,
- seeking solutions that will bring advantage to all those who will be affected,
- deciding on the actions which will be taken and who, in accordance with this contract, will take them and
- deciding which risks have now been avoided or have passed and can be removed from the Risk Register.

16.4 The *Project Manager* revises the Risk Register to record the decisions made at each risk reduction meeting and issues the revised Risk Register to the *Contractor*. If a decision needs a change to the Works Information, the *Project Manager* instructs the change at the same time as he issues the revised Risk Register.

Ambiguities and inconsistencies **17**

17.1 The *Project Manager* or the *Contractor* notifies the other as soon as either becomes aware of an ambiguity or inconsistency in or between the documents which are part of this contract. The *Project Manager* gives an instruction resolving the ambiguity or inconsistency.

Illegal and impossible requirements **18**

18.1 The *Contractor* notifies the *Project Manager* as soon as he considers that the Works Information requires him to do anything which is illegal or impossible. If the *Project Manager* agrees, he gives an instruction to change the Works Information appropriately.

Prevention **19**

19.1 If an event occurs which

- stops the *Contractor* completing the *works* or
- stops the *Contractor* completing the *works* by the date shown on the Accepted Programme,

and which

- neither Party could prevent and
- an experienced contractor would have judged at the Contract Date to have such a small chance of occurring that it would have been unreasonable for him to have allowed for it,

the *Project Manager* gives an instruction to the *Contractor* stating how he is to deal with the event.

2 The *Contractor*'s main responsibilities

Providing the Works **20**

20.1 The *Contractor* Provides the Works in accordance with the Works Information.

20.3 The *Contractor* advises the *Project Manager* on the practical implications of the design of the *works* and on subcontracting arrangements.

20.4 The *Contractor* prepares forecasts of the total Defined Cost for the whole of the *works* in consultation with the *Project Manager* and submits them to the *Project Manager*. Forecasts are prepared at the intervals stated in the Contract Data from the *starting date* until Completion of the whole of the *works*. An explanation of the changes made since the previous forecast is submitted with each forecast.

The *Contractor*'s design **21**

21.1 The *Contractor* designs the parts of the *works* which the Works Information states he is to design.

21.2 The *Contractor* submits the particulars of his design as the Works Information requires to the *Project Manager* for acceptance. A reason for not accepting the *Contractor*'s design is that it does not comply with either the Works Information or the applicable law.

The *Contractor* does not proceed with the relevant work until the *Project Manager* has accepted his design.

21.3 The *Contractor* may submit his design for acceptance in parts if the design of each part can be assessed fully.

Using the *Contractor*'s design **22**

22.1 The *Employer* may use and copy the *Contractor*'s design for any purpose connected with construction, use, alteration or demolition of the *works* unless otherwise stated in the Works Information and for other purposes as stated in the Works Information.

Design of Equipment **23**

23.1 The *Contractor* submits particulars of the design of an item of Equipment to the *Project Manager* for acceptance if the *Project Manager* instructs him to. A reason for not accepting is that the design of the item will not allow the *Contractor* to Provide the Works in accordance with

- the Works Information,
- the *Contractor*'s design which the *Project Manager* has accepted or
- the applicable law.

People **24**

24.1 The *Contractor* either employs each key person named to do the job stated in the Contract Data or employs a replacement person who has been accepted by the *Project Manager*. The *Contractor* submits the name, relevant qualifications and experience of a proposed replacement person to the *Project Manager* for acceptance. A reason for not accepting the person is that his relevant qualifications and experience are not as good as those of the person who is to be replaced.

24.2 The *Project Manager* may, having stated his reasons, instruct the *Contractor* to remove an employee. The *Contractor* then arranges that, after one day, the employee has no further connection with the work included in this contract.

Working with the *Employer* and Others **25**

25.1 The *Contractor* co-operates with Others in obtaining and providing information which they need in connection with the *works*. He co-operates with Others and shares the Working Areas with them as stated in the Works Information.

25.2 The *Employer* and the *Contractor* provide services and other things as stated in the Works Information. Any cost incurred by the *Employer* as a result of the *Contractor* not providing the services and other things which he is to provide is assessed by the *Project Manager* and paid by the *Contractor*.

25.3 If the *Project Manager* decides that the work does not meet the Condition stated for a Key Date by the date stated and, as a result, the *Employer* incurs additional cost either

- in carrying out work or
- by paying an additional amount to Others in carrying out work

on the same project, the additional cost which the *Employer* has paid or will incur is paid by the *Contractor*. The *Project Manager* assesses the additional cost within four weeks of the date when the Condition for the Key Date is met. The *Employer*'s right to recover the additional cost is his only right in these circumstances.

Subcontracting **26**

26.1 If the *Contractor* subcontracts work, he is responsible for Providing the Works as if he had not subcontracted. This contract applies as if a Subcontractor's employees and equipment were the *Contractor*'s.

26.2 The *Contractor* submits the name of each proposed Subcontractor to the *Project Manager* for acceptance. A reason for not accepting the Subcontractor is that his appointment will not allow the *Contractor* to Provide the Works. The *Contractor* does not appoint a proposed Subcontractor until the *Project Manager* has accepted him.

26.3 The *Contractor* submits the proposed conditions of contract for each subcontract to the *Project Manager* for acceptance unless

- an NEC contract is proposed or
- the *Project Manager* has agreed that no submission is required.

The *Contractor* does not appoint a Subcontractor on the proposed subcontract conditions submitted until the *Project Manager* has accepted them. A reason for not accepting them is that

- they will not allow the *Contractor* to Provide the Works or
- they do not include a statement that the parties to the subcontract shall act in a spirit of mutual trust and co-operation.

26.4 The *Contractor* submits the proposed contract data for each subcontract for acceptance to the *Project Manager* if

- **an NEC contract is proposed and**
- **the *Project Manager* instructs the *Contractor* to make the submission.**

A reason for not accepting the proposed contract data is that its use will not allow the *Contractor* to Provide the Works.

Other responsibilities **27**

27.1 The *Contractor* obtains approval of his design from Others where necessary.

27.2 The *Contractor* provides access to work being done and to Plant and Materials being stored for this contract for

- the *Project Manager*,
- the *Supervisor* and
- Others notified to him by the *Project Manager*.

27.3 The *Contractor* obeys an instruction which is in accordance with this contract and is given to him by the *Project Manager* or the *Supervisor*.

27.4 The *Contractor* acts in accordance with the health and safety requirements stated in the Works Information.

3 Time

Starting, Completion and Key Dates

30

30.1 The *Contractor* does not start work on the Site until the first *access date* and does the work so that Completion is on or before the Completion Date.

30.2 The *Project Manager* decides the date of Completion. The *Project Manager* certifies Completion within one week of Completion.

30.3 The *Contractor* does the work so that the Condition stated for each Key Date is met by the Key Date.

The programme

31

31.1 If a programme is not identified in the Contract Data, the *Contractor* submits a first programme to the *Project Manager* for acceptance within the period stated in the Contract Data.

31.2 The *Contractor* shows on each programme which he submits for acceptance

- the *starting date, access dates*, Key Dates and Completion Date,
- planned Completion,
- the order and timing of the operations which the *Contractor* plans to do in order to Provide the Works,
- the order and timing of the work of the *Employer* and Others as last agreed with them by the *Contractor* or, if not so agreed, as stated in the Works Information,
- the dates when the *Contractor* plans to meet each Condition stated for the Key Dates and to complete other work needed to allow the *Employer* and Others to do their work,
- provisions for
  - float,
  - time risk allowances,
  - health and safety requirements and
  - the procedures set out in this contract,
- the dates when, in order to Provide the Works in accordance with his programme, the *Contractor* will need
  - access to a part of the Site if later than its *access date*,
  - acceptances,
  - Plant and Materials and other things to be provided by the *Employer* and
  - information from Others,
- for each operation, a statement of how the *Contractor* plans to do the work identifying the principal Equipment and other resources which he plans to use and
- other information which the Works Information requires the *Contractor* to show on a programme submitted for acceptance.

31.3 Within two weeks of the *Contractor* submitting a programme to him for acceptance, the *Project Manager* either accepts the programme or notifies the *Contractor* of his reasons for not accepting it. A reason for not accepting a programme is that

- the *Contractor*'s plans which it shows are not practicable,
- it does not show the information which this contract requires,
- it does not represent the *Contractor*'s plans realistically or
- it does not comply with the Works Information.

Revising the programme **32**

32.1 The *Contractor* shows on each revised programme

- the actual progress achieved on each operation and its effect upon the timing of the remaining work,
- the effects of implemented compensation events,
- how the *Contractor* plans to deal with any delays and to correct notified Defects and
- any other changes which the *Contractor* proposes to make to the Accepted Programme.

32.2 The *Contractor* submits a revised programme to the *Project Manager* for acceptance

- within the *period for reply* after the *Project Manager* has instructed him to,
- when the *Contractor* chooses to and, in any case,
- at no longer interval than the interval stated in the Contract Data from the *starting date* until Completion of the whole of the *works*.

Access to and use of the Site **33**

33.1 The *Employer* allows access to and use of each part of the Site to the *Contractor* which is necessary for the work included in this contract. Access and use is allowed on or before the later of its *access date* and the date for access shown on the Accepted Programme.

Instructions to stop or not to start work **34**

34.1 The *Project Manager* may instruct the *Contractor* to stop or not to start any work and may later instruct him that he may re-start or start it.

Take over **35**

35.1 The *Employer* need not take over the *works* before the Completion Date if it is stated in the Contract Data that he is not willing to do so. Otherwise the *Employer* takes over the *works* not later than two weeks after Completion.

35.2 The *Employer* may use any part of the *works* before Completion has been certified. If he does so, he takes over the part of the *works* when he begins to use it except if the use is

- for a reason stated in the Works Information or
- to suit the *Contractor*'s method of working.

35.3 The *Project Manager* certifies the date upon which the *Employer* takes over any part of the *works* and its extent within one week of the date.

Acceleration **36**

36.1 The *Project Manager* may instruct the *Contractor* to submit a quotation for an acceleration to achieve Completion before the Completion Date. The *Project Manager* states changes to the Key Dates to be included in the quotation. A quotation for an acceleration comprises proposed changes to the Prices and a revised programme showing the earlier Completion Date and the changed Key Dates. The *Contractor* submits details of his assessment with each quotation.

36.2 The *Contractor* submits a quotation or gives his reasons for not doing so within the *period for reply*.

36.4 **When the *Project Manager* accepts a quotation for an acceleration, he changes the Completion Date, the Key Dates and the forecast of the total Defined Cost of the whole of the *works* accordingly and accepts the revised programme.**

core clauses | main option clauses | secondary option clauses | cost components | contract data

4 Testing and Defects

Tests and inspections 40

40.1 The subclauses in this clause only apply to tests and inspections required by the Works Information or the applicable law.

40.2 The *Contractor* and the *Employer* provide materials, facilities and samples for tests and inspections as stated in the Works Information.

40.3 The *Contractor* and the *Supervisor* each notifies the other of each of his tests and inspections before it starts and afterwards notifies the other of its results. The *Contractor* notifies the *Supervisor* in time for a test or inspection to be arranged and done before doing work which would obstruct the test or inspection. The *Supervisor* may watch any test done by the *Contractor*.

40.4 If a test or inspection shows that any work has a Defect, the *Contractor* corrects the Defect and the test or inspection is repeated.

40.5 The *Supervisor* does his tests and inspections without causing unnecessary delay to the work or to a payment which is conditional upon a test or inspection being successful. A payment which is conditional upon a *Supervisor*'s test or inspection being successful becomes due at the later of the *defects date* and the end of the last *defect correction period* if

- the *Supervisor* has not done the test or inspection and
- the delay to the test or inspection is not the *Contractor*'s fault.

40.6 The *Project Manager* assesses the cost incurred by the *Employer* in repeating a test or inspection after a Defect is found. The *Contractor* pays the amount assessed.

40.7 When the *Project Manager* assesses the cost incurred by the *Employer* in repeating a test or inspection after a Defect is found, the *Project Manager* does not include the *Contractor*'s cost of carrying out the repeat test or inspection.

Testing and inspection before delivery 41

41.1 The *Contractor* does not bring to the Working Areas those Plant and Materials which the Works Information states are to be tested or inspected before delivery until the *Supervisor* has notified the *Contractor* that they have passed the test or inspection.

Searching for and notifying Defects 42

42.1 Until the *defects date*, the *Supervisor* may instruct the *Contractor* to search for a Defect. He gives his reason for the search with his instruction. Searching may include

- uncovering, dismantling, re-covering and re-erecting work,
- providing facilities, materials and samples for tests and inspections done by the *Supervisor* and
- doing tests and inspections which the Works Information does not require.

42.2 Until the *defects date*, the *Supervisor* notifies the *Contractor* of each Defect as soon as he finds it and the *Contractor* notifies the *Supervisor* of each Defect as soon as he finds it.

Correcting Defects 43

43.1 The *Contractor* corrects a Defect whether or not the *Supervisor* notifies him of it.

43.2 The *Contractor* corrects a notified Defect before the end of the *defect correction period*. The *defect correction period* begins at Completion for Defects notified before Completion and when the Defect is notified for other Defects.

43.3 The *Supervisor* issues the Defects Certificate at the later of the *defects date* and the end of the last *defect correction period*. The *Employer's* rights in respect of a Defect which the *Supervisor* has not found or notified are not affected by the issue of the Defects Certificate.

43.4 The *Project Manager* arranges for the *Employer* to allow the *Contractor* access to and use of a part of the *works* which he has taken over if they are needed for correcting a Defect. In this case the *defect correction period* begins when the necessary access and use have been provided.

Accepting Defects 44

44.1 The *Contractor* and the *Project Manager* may each propose to the other that the Works Information should be changed so that a Defect does not have to be corrected.

44.2 If the *Contractor* and the *Project Manager* are prepared to consider the change, the *Contractor* submits a quotation for reduced Prices or an earlier Completion Date or both to the *Project Manager* for acceptance. If the *Project Manager* accepts the quotation, he gives an instruction to change the Works Information, the Prices and the Completion Date accordingly.

Uncorrected Defects 45

45.1 If the *Contractor* is given access in order to correct a notified Defect but he has not corrected it within its *defect correction period*, the *Project Manager* assesses the cost to the *Employer* of having the Defect corrected by other people and the *Contractor* pays this amount. The Works Information is treated as having been changed to accept the Defect.

45.2 If the *Contractor* is not given access in order to correct a notified Defect before the *defects date*, the *Project Manager* assesses the cost to the *Contractor* of correcting the Defect and the *Contractor* pays this amount. The Works Information is treated as having been changed to accept the Defect.

5 Payment

Assessing the amount due **50**

50.1 The *Project Manager* assesses the amount due at each assessment date. The first assessment date is decided by the *Project Manager* to suit the procedures of the Parties and is not later than the *assessment interval* after the *starting date*. Later assessment dates occur

- at the end of each *assessment interval* until four weeks after the *Supervisor* issues the Defects Certificate and
- at Completion of the whole of the *works*.

50.2 The amount due is

- the Price for Work Done to Date,
- plus other amounts to be paid to the *Contractor*,
- less amounts to be paid by or retained from the *Contractor*.

Any tax which the law requires the *Employer* to pay to the *Contractor* is included in the amount due.

50.3 If no programme is identified in the Contract Data, one quarter of the Price for Work Done to Date is retained in assessments of the amount due until the *Contractor* has submitted a first programme to the *Project Manager* for acceptance showing the information which this contract requires.

50.4 In assessing the amount due, the *Project Manager* considers any application for payment the *Contractor* has submitted on or before the assessment date. The *Project Manager* gives the *Contractor* details of how the amount due has been assessed.

50.5 The *Project Manager* corrects any wrongly assessed amount due in a later payment certificate.

50.7 **Payments of Defined Cost made by the *Contractor* in a currency other than the *currency of this contract* are included in the amount due as payments to be made to him in the same currency. Such payments are converted to the *currency of this contract* in order to calculate the Fee using the *exchange rates*.**

Payment **51**

51.1 The *Project Manager* certifies a payment within one week of each assessment date. The first payment is the amount due. Other payments are the change in the amount due since the last payment certificate. A payment is made by the *Contractor* to the *Employer* if the change reduces the amount due. Other payments are made by the *Employer* to the *Contractor*. Payments are in the *currency of this contract* unless otherwise stated in this contract.

51.2 Each certified payment is made within three weeks of the assessment date or, if a different period is stated in the Contract Data, within the period stated. If a certified payment is late, or if a payment is late because the *Project Manager* does not issue a certificate which he should issue, interest is paid on the late payment. Interest is assessed from the date by which the late payment should have been made until the date when the late payment is made, and is included in the first assessment after the late payment is made.

51.3 If an amount due is corrected in a later certificate either

- by the *Project Manager* in relation to a mistake or a compensation event or
- following a decision of the *Adjudicator* or the *tribunal*,

interest on the correcting amount is paid. Interest is assessed from the date when the incorrect amount was certified until the date when the correcting amount is certified and is included in the assessment which includes the correcting amount.

51.4 Interest is calculated on a daily basis at the *interest rate* and is compounded annually.

Defined Cost **52**

52.1 All the *Contractor*'s costs which are not included in the Defined Cost are treated as included in the Fee. Defined Cost includes only amounts calculated using rates and percentages stated in the Contract Data and other amounts at open market or competitively tendered prices with deductions for all discounts, rebates and taxes which can be recovered.

52.2 The *Contractor* keeps these records

- **accounts of payments of Defined Cost,**
- **proof that the payments have been made,**
- **communications about and assessments of compensation events for Subcontractors and**
- **other records as stated in the Works Information.**

52.3 The *Contractor* allows the *Project Manager* to inspect at any time within working hours the accounts and records which he is required to keep.

6 Compensation events

Compensation events **60**

60.1 The following are compensation events.

(1) The *Project Manager* gives an instruction changing the Works Information except

- a change made in order to accept a Defect or
- a change to the Works Information provided by the *Contractor* for his design which is made either at his request or to comply with other Works Information provided by the *Employer*.

(2) The *Employer* does not allow access to and use of a part of the Site by the later of its *access date* and the date shown on the Accepted Programme.

(3) The *Employer* does not provide something which he is to provide by the date for providing it shown on the Accepted Programme.

(4) The *Project Manager* gives an instruction to stop or not to start any work or to change a Key Date.

(5) The *Employer* or Others

- do not work within the times shown on the Accepted Programme,
- do not work within the conditions stated in the Works Information or
- carry out work on the Site that is not stated in the Works Information.

(6) The *Project Manager* or the *Supervisor* does not reply to a communication from the *Contractor* within the period required by this contract.

(7) The *Project Manager* gives an instruction for dealing with an object of value or of historical or other interest found within the Site.

(8) The *Project Manager* or the *Supervisor* changes a decision which he has previously communicated to the *Contractor.*

(9) The *Project Manager* withholds an acceptance (other than acceptance of a quotation for acceleration or for not correcting a Defect) for a reason not stated in this contract.

(10) The *Supervisor* instructs the *Contractor* to search for a Defect and no Defect is found unless the search is needed only because the *Contractor* gave insufficient notice of doing work obstructing a required test or inspection.

(11) A test or inspection done by the *Supervisor* causes unnecessary delay.

(12) The *Contractor* encounters physical conditions which

- are within the Site,
- are not weather conditions and
- an experienced contractor would have judged at the Contract Date to have such a small chance of occurring that it would have been unreasonable for him to have allowed for them.

Only the difference between the physical conditions encountered and those for which it would have been reasonable to have allowed is taken into account in assessing a compensation event.

(13) A *weather measurement* is recorded

- within a calendar month,
- before the Completion Date for the whole of the *works* and
- at the place stated in the Contract Data

the value of which, by comparison with the *weather data*, is shown to occur on average less frequently than once in ten years.

Only the difference between the *weather measurement* and the weather which the *weather data* show to occur on average less frequently than once in ten years is taken into account in assessing a compensation event.

(14) An event which is an *Employer*'s risk stated in this contract.

(15) The *Project Manager* certifies take over of a part of the *works* before both Completion and the Completion Date.

(16) The *Employer* does not provide materials, facilities and samples for tests and inspections as stated in the Works Information.

(17) The *Project Manager* notifies a correction to an assumption which he has stated about a compensation event.

(18) A breach of contract by the *Employer* which is not one of the other compensation events in this contract.

(19) An event which

- stops the *Contractor* completing the *works* or
- stops the *Contractor* completing the *works* by the date shown on the Accepted Programme,

and which

- neither Party could prevent,
- an experienced contractor would have judged at the Contract Date to have such a small chance of occurring that it would have been unreasonable for him to have allowed for it and
- is not one of the other compensation events stated in this contract.

60.2 In judging the physical conditions for the purpose of assessing a compensation event, the *Contractor* is assumed to have taken into account

- the Site Information,
- publicly available information referred to in the Site Information,
- information obtainable from a visual inspection of the Site and
- other information which an experienced contractor could reasonably be expected to have or to obtain.

60.3 If there is an ambiguity or inconsistency within the Site Information (including the information referred to in it), the *Contractor* is assumed to have taken into account the physical conditions more favourable to doing the work.

Notifying compensation events **61**

61.1 For compensation events which arise from the *Project Manager* or the *Supervisor* giving an instruction or changing an earlier decision, the *Project Manager* notifies the *Contractor* of the compensation event at the time of giving the instruction or changing the earlier decision. He also instructs the *Contractor* to submit quotations, unless the event arises from a fault of the *Contractor* or quotations have already been submitted. The *Contractor* puts the instruction or changed decision into effect.

61.2 The *Project Manager* may instruct the *Contractor* to submit quotations for a proposed instruction or a proposed changed decision. The *Contractor* does not put a proposed instruction or a proposed changed decision into effect.

61.3 The *Contractor* notifies the *Project Manager* of an event which has happened or which he expects to happen as a compensation event if

- the *Contractor* believes that the event is a compensation event and
- the *Project Manager* has not notified the event to the *Contractor*.

If the *Contractor* does not notify a compensation event within eight weeks of becoming aware of the event, he is not entitled to a change in the Prices, the Completion Date or a Key Date unless the *Project Manager* should have notified the event to the *Contractor* but did not.

61.4 If the *Project Manager* decides that an event notified by the *Contractor*

- arises from a fault of the *Contractor*,
- has not happened and is not expected to happen,
- has no effect upon Defined Cost, Completion or meeting a Key Date or
- is not one of the compensation events stated in this contract

he notifies the *Contractor* of his decision that the Prices, the Completion Date and the Key Dates are not to be changed.

If the *Project Manager* decides otherwise, he notifies the *Contractor* accordingly and instructs him to submit quotations.

If the *Project Manager* does not notify his decision to the *Contractor* within either

- one week of the *Contractor*'s notification or
- a longer period to which the *Contractor* has agreed,

the *Contractor* may notify the *Project Manager* to this effect. A failure by the *Project Manager* to reply within two weeks of this notification is treated as acceptance by the *Project Manager* that the event is a compensation event and an instruction to submit quotations.

61.5 If the *Project Manager* decides that the *Contractor* did not give an early warning of the event which an experienced contractor could have given, he notifies this decision to the *Contractor* when he instructs him to submit quotations.

61.6 If the *Project Manager* decides that the effects of a compensation event are too uncertain to be forecast reasonably, he states assumptions about the event in his instruction to the *Contractor* to submit quotations. Assessment of the event is based on these assumptions. If any of them is later found to have been wrong, the *Project Manager* notifies a correction.

61.7 A compensation event is not notified after the *defects date*.

Quotations for compensation events **62**

62.1 After discussing with the *Contractor* different ways of dealing with the compensation event which are practicable, the *Project Manager* may instruct the *Contractor* to submit alternative quotations. The *Contractor* submits the required quotations to the *Project Manager* and may submit quotations for other methods of dealing with the compensation event which he considers practicable.

62.2 Quotations for compensation events comprise proposed changes to the Prices and any delay to the Completion Date and Key Dates assessed by the *Contractor*. The *Contractor* submits details of his assessment with each quotation. If the programme for remaining work is altered by the compensation event, the *Contractor* includes the alterations to the Accepted Programme in his quotation.

62.3 The *Contractor* submits quotations within three weeks of being instructed to do so by the *Project Manager*. The *Project Manager* replies within two weeks of the submission. His reply is

- an instruction to submit a revised quotation,
- an acceptance of a quotation,
- a notification that a proposed instruction will not be given or a proposed changed decision will not be made or
- a notification that he will be making his own assessment.

62.4 The *Project Manager* instructs the *Contractor* to submit a revised quotation only after explaining his reasons for doing so to the *Contractor*. The *Contractor* submits the revised quotation within three weeks of being instructed to do so.

62.5 The *Project Manager* extends the time allowed for

- the *Contractor* to submit quotations for a compensation event and
- the *Project Manager* to reply to a quotation

if the *Project Manager* and the *Contractor* agree to the extension before the submission or reply is due. The *Project Manager* notifies the extension that has been agreed to the *Contractor*.

62.6 If the *Project Manager* does not reply to a quotation within the time allowed, the *Contractor* may notify the *Project Manager* to this effect. If the *Contractor* submitted more than one quotation for the compensation event, he states in his notification which quotation he proposes is to be accepted. If the *Project Manager* does not reply to the notification within two weeks, and unless the quotation is for a proposed instruction or a proposed changed decision, the *Contractor*'s notification is treated as acceptance of the quotation by the *Project Manager*.

Assessing compensation events **63**

63.1 The changes to the Prices are assessed as the effect of the compensation event upon

- the actual Defined Cost of the work already done,
- the forecast Defined Cost of the work not yet done and
- the resulting Fee.

The date when the *Project Manager* instructed or should have instructed the *Contractor* to submit quotations divides the work already done from the work not yet done.

63.2 If the effect of a compensation event is to reduce the total Defined Cost, the Prices are not reduced except as stated in this contract.

63.3 A delay to the Completion Date is assessed as the length of time that, due to the compensation event, planned Completion is later than planned Completion as shown on the Accepted Programme. A delay to a Key Date is assessed as the length of time that, due to the compensation event, the planned date when the Condition stated for a Key Date will be met is later than the date shown on the Accepted Programme.

63.4 The rights of the *Employer* and the *Contractor* to changes to the Prices, the Completion Date and the Key Dates are their only rights in respect of a compensation event.

63.5 If the *Project Manager* has notified the *Contractor* of his decision that the *Contractor* did not give an early warning of a compensation event which an experienced contractor could have given, the event is assessed as if the *Contractor* had given early warning.

63.6 Assessment of the effect of a compensation event includes risk allowances for cost and time for matters which have a significant chance of occurring and are at the *Contractor*'s risk under this contract.

63.7 Assessments are based upon the assumptions that the *Contractor* reacts competently and promptly to the compensation event, that any Defined Cost and time due to the event are reasonably incurred and that the Accepted Programme can be changed.

63.8 A compensation event which is an instruction to change the Works Information in order to resolve an ambiguity or inconsistency is assessed as if the Prices, the Completion Date and the Key Dates were for the interpretation most favourable to the Party which did not provide the Works Information.

63.9 If a change to the Works Information makes the description of the Condition for a Key Date incorrect, the *Project Manager* corrects the description. This correction is taken into account in assessing the compensation event for the change to the Works Information.

63.15 If the *Project Manager* and the *Contractor* agree, the *Contractor* assesses a compensation event using the Shorter Schedule of Cost Components. The *Project Manager* may make his own assessments using the Shorter Schedule of Cost Components.

The *Project Manager*'s assessments

64

64.1 The *Project Manager* assesses a compensation event

- if the *Contractor* has not submitted a quotation and details of his assessment within the time allowed,
- if the *Project Manager* decides that the *Contractor* has not assessed the compensation event correctly in a quotation and he does not instruct the *Contractor* to submit a revised quotation,
- if, when the *Contractor* submits quotations for a compensation event, he has not submitted a programme or alterations to a programme which this contract requires him to submit or
- if, when the *Contractor* submits quotations for a compensation event, the *Project Manager* has not accepted the *Contractor*'s latest programme for one of the reasons stated in this contract.

64.2 The *Project Manager* assesses a compensation event using his own assessment of the programme for the remaining work if

- there is no Accepted Programme or
- the *Contractor* has not submitted a programme or alterations to a programme for acceptance as required by this contract.

64.3 The *Project Manager* notifies the *Contractor* of his assessment of a compensation event and gives him details of it within the period allowed for the *Contractor*'s submission of his quotation for the same event. This period starts when the need for the *Project Manager*'s assessment becomes apparent.

64.4 If the *Project Manager* does not assess a compensation event within the time allowed, the *Contractor* may notify the *Project Manager* to this effect. If the *Contractor* submitted more than one quotation for the compensation event, he states in his notification which quotation he proposes is to be accepted. If the *Project Manager* does not reply within two weeks of this notification the notification is treated as acceptance of the *Contractor's* quotation by the *Project Manager*.

Implementing compensation events

65

65.1 A compensation event is implemented when

- the *Project Manager* notifies his acceptance of the *Contractor*'s quotation,
- the *Project Manager* notifies the *Contractor* of his own assessment or
- a *Contractor*'s quotation is treated as having been accepted by the *Project Manager*.

65.2 The assessment of a compensation event is not revised if a forecast upon which it is based is shown by later recorded information to have been wrong.

65.3 The changes to the forecast amount of the Prices, the Completion Date and the Key Dates are included in the notification implementing a compensation event.

7 Title

The *Employer*'s title to Plant and Materials

70

70.1 Whatever title the *Contractor* has to Plant and Materials which is outside the Working Areas passes to the *Employer* if the *Supervisor* has marked it as for this contract.

70.2 Whatever title the *Contractor* has to Plant and Materials passes to the *Employer* if it has been brought within the Working Areas. The title to Plant and Materials passes back to the *Contractor* if it is removed from the Working Areas with the *Project Manager*'s permission.

Marking Equipment, Plant and Materials outside the Working Areas

71

71.1 The *Supervisor* marks Equipment, Plant and Materials which are outside the Working Areas if

- this contract identifies them for payment and
- the *Contractor* has prepared them for marking as the Works Information requires.

Removing Equipment

72

72.1 The *Contractor* removes Equipment from the Site when it is no longer needed unless the *Project Manager* allows it to be left in the *works*.

Objects and materials within the Site

73

73.1 The *Contractor* has no title to an object of value or of historical or other interest within the Site. The *Contractor* notifies the *Project Manager* when such an object is found and the *Project Manager* instructs the *Contractor* how to deal with it. The *Contractor* does not move the object without instructions.

73.2 The *Contractor* has title to materials from excavation and demolition only as stated in the Works Information.

8 Risks and insurance

Employer's risks 80

80.1 The following are *Employer*'s risks.

- Claims, proceedings, compensation and costs payable which are due to
  - use or occupation of the Site by the *works* or for the purpose of the *works* which is the unavoidable result of the *works*,
  - negligence, breach of statutory duty or interference with any legal right by the *Employer* or by any person employed by or contracted to him except the *Contractor* or
  - a fault of the *Employer* or a fault in his design.
- Loss of or damage to Plant and Materials supplied to the *Contractor* by the *Employer*, or by Others on the *Employer*'s behalf, until the *Contractor* has received and accepted them.
- Loss of or damage to the *works*, Plant and Materials due to
  - war, civil war, rebellion, revolution, insurrection, military or usurped power,
  - strikes, riots and civil commotion not confined to the *Contractor*'s employees or
  - radioactive contamination.
- Loss of or wear or damage to the parts of the *works* taken over by the *Employer*, except loss, wear or damage occurring before the issue of the Defects Certificate which is due to
  - a Defect which existed at take over,
  - an event occurring before take over which was not itself an *Employer*'s risk or
  - the activities of the *Contractor* on the Site after take over.
- Loss of or wear or damage to the *works* and any Equipment, Plant and Materials retained on the Site by the *Employer* after a termination, except loss, wear or damage due to the activities of the *Contractor* on the Site after the termination.
- Additional *Employer*'s risks stated in the Contract Data.

The *Contractor*'s risks 81

81.1 From the *starting date* until the Defects Certificate has been issued, the risks which are not carried by the *Employer* are carried by the *Contractor*.

Repairs 82

82.1 Until the Defects Certificate has been issued and unless otherwise instructed by the *Project Manager,* the *Contractor* promptly replaces loss of and repairs damage to the *works*, Plant and Materials.

Indemnity 83

83.1 Each Party indemnifies the other against claims, proceedings, compensation and costs due to an event which is at his risk.

83.2 The liability of each Party to indemnify the other is reduced if events at the other Party's risk contributed to the claims, proceedings, compensation and costs. The reduction is in proportion to the extent that events which were at the other Party's risk contributed, taking into account each Party's responsibilities under this contract.

Insurance cover 84

84.1 The *Contractor* provides the insurances stated in the Insurance Table except any insurance which the *Employer* is to provide as stated in the Contract Data. The *Contractor* provides additional insurances as stated in the Contract Data.

84.2 The insurances are in the joint names of the Parties and provide cover for events which are at the *Contractor*'s risk from the *starting date* until the Defects Certificate or a termination certificate has been issued.

INSURANCE TABLE

Insurance against	Minimum amount of cover or minimum limit of indemnity
Loss of or damage to the *works*, Plant and Materials	The replacement cost, including the amount stated in the Contract Data for the replacement of any Plant and Materials provided by the *Employer*
Loss of or damage to Equipment	The replacement cost
Liability for loss of or damage to property (except the *works*, Plant and Materials and Equipment) and liability for bodily injury to or death of a person (not an employee of the *Contractor*) caused by activity in connection with this contract	The amount stated in the Contract Data for any one event with cross liability so that the insurance applies to the Parties separately
Liability for death of or bodily injury to employees of the *Contractor* arising out of and in the course of their employment in connection with this contract	The greater of the amount required by the applicable law and the amount stated in the Contract Data for any one event

Insurance policies 85

85.1 Before the *starting date* and on each renewal of the insurance policy until the *defects date*, the *Contractor* submits to the *Project Manager* for acceptance certificates which state that the insurance required by this contract is in force. The certificates are signed by the *Contractor*'s insurer or insurance broker. A reason for not accepting the certificates is that they do not comply with this contract.

85.2 Insurance policies include a waiver by the insurers of their subrogation rights against directors and other employees of every insured except where there is fraud.

85.3 The Parties comply with the terms and conditions of the insurance policies.

85.4 Any amount not recovered from an insurer is borne by the *Employer* for events which are at his risk and by the *Contractor* for events which are at his risk.

If the *Contractor* does not insure 86

86.1 The *Employer* may insure a risk which this contract requires the *Contractor* to insure if the *Contractor* does not submit a required certificate. The cost of this insurance to the *Employer* is paid by the *Contractor*.

Insurance by the *Employer* 87

87.1 The *Project Manager* submits policies and certificates for insurances provided by the *Employer* to the *Contractor* for acceptance before the *starting date* and afterwards as the *Contractor* instructs. The *Contractor* accepts the policies and certificates if they comply with this contract.

core clauses | main option clauses | secondary option clauses | cost components | contract data

87.2 The *Contractor*'s acceptance of an insurance policy or certificate provided by the *Employer* does not change the responsibility of the *Employer* to provide the insurances stated in the Contract Data.

87.3 The *Contractor* may insure a risk which this contract requires the *Employer* to insure if the *Employer* does not submit a required policy or certificate. The cost of this insurance to the *Contractor* is paid by the *Employer*.

9 Termination

Termination **90**

90.1 If either Party wishes to terminate the *Contractor*'s obligation to Provide the Works he notifies the *Project Manager* and the other Party giving details of his reason for terminating. The *Project Manager* issues a termination certificate to both Parties promptly if the reason complies with this contract.

90.2 The *Contractor* may terminate only for a reason identified in the Termination Table. The *Employer* may terminate for any reason. The procedures followed and the amounts due on termination are in accordance with the Termination Table.

TERMINATION TABLE

Terminating Party	Reason	Procedure	Amount due
The *Employer*	A reason other than R1–R21	P1 and P2	A1, A2 and A4
	R1–R15 or R18	P1, P2 and P3	A1 and A3
	R17 or R20	P1 and P3	A1 and A2
	R21	P1 and P4	A1 and A2
The *Contractor*	R1–R10, R16 or R19	P1 and P4	A1, A2 and A4
	R17 or R20	P1 and P4	A1 and A2

90.3 The procedures for termination are implemented immediately after the *Project Manager* has issued a termination certificate.

90.4 Within thirteen weeks of termination, the *Project Manager* certifies a final payment to or from the *Contractor* which is the *Project Manager*'s assessment of the amount due on termination less the total of previous payments. Payment is made within three weeks of the *Project Manager*'s certificate.

90.5 After a termination certificate has been issued, the *Contractor* does no further work necessary to Provide the Works.

Reasons for termination **91**

91.1 Either Party may terminate if the other Party has done one of the following or its equivalent.

- If the other Party is an individual and has
  - presented his petition for bankruptcy (R1),
  - had a bankruptcy order made against him (R2),
  - had a receiver appointed over his assets (R3) or
  - made an arrangement with his creditors (R4).
- If the other Party is a company or partnership and has
  - had a winding-up order made against it (R5),
  - had a provisional liquidator appointed to it (R6),
  - passed a resolution for winding-up (other than in order to amalgamate or reconstruct) (R7),
  - had an administration order made against it (R8),
  - had a receiver, receiver and manager, or administrative receiver appointed over the whole or a substantial part of its undertaking or assets (R9) or
  - made an arrangement with its creditors (R10).

91.2 The *Employer* may terminate if the *Project Manager* has notified that the *Contractor* has defaulted in one of the following ways and not put the default right within four weeks of the notification.

- Substantially failed to comply with his obligations (R11).
- Not provided a bond or guarantee which this contract requires (R12).
- Appointed a Subcontractor for substantial work before the *Project Manager* has accepted the Subcontractor (R13).

91.3 The *Employer* may terminate if the *Project Manager* has notified that the *Contractor* has defaulted in one of the following ways and not stopped defaulting within four weeks of the notification.

- Substantially hindered the *Employer* or Others (R14).
- Substantially broken a health or safety regulation (R15).

91.4 The *Contractor* may terminate if the *Employer* has not paid an amount certified by the *Project Manager* within thirteen weeks of the date of the certificate (R16).

91.5 Either Party may terminate if the Parties have been released under the law from further performance of the whole of this contract (R17).

91.6 If the *Project Manager* has instructed the *Contractor* to stop or not to start any substantial work or all work and an instruction allowing the work to re-start or start has not been given within thirteen weeks,

- the *Employer* may terminate if the instruction was due to a default by the *Contractor* (R18),
- the *Contractor* may terminate if the instruction was due to a default by the *Employer* (R19) and
- either Party may terminate if the instruction was due to any other reason (R20).

91.7 The *Employer* may terminate if an event occurs which

- stops the *Contractor* completing the *works* or
- stops the *Contractor* completing the *works* by the date shown on the Accepted Programme and is forecast to delay Completion by more than 13 weeks,

and which

- neither Party could prevent and
- an experienced contractor would have judged at the Contract Date to have such a small chance of occurring that it would have been unreasonable for him to have allowed for it (R21).

Procedures on termination **92**

92.1 On termination, the *Employer* may complete the *works* and may use any Plant and Materials to which he has title (P1).

92.2 The procedure on termination also includes one or more of the following as set out in the Termination Table.

P2 The *Employer* may instruct the *Contractor* to leave the Site, remove any Equipment, Plant and Materials from the Site and assign the benefit of any subcontract or other contract related to performance of this contract to the *Employer*.

P3 The *Employer* may use any Equipment to which the *Contractor* has title to complete the *works*. The *Contractor* promptly removes the Equipment from Site when the *Project Manager* notifies him that the *Employer* no longer requires it to complete the *works*.

P4 The *Contractor* leaves the Working Areas and removes the Equipment.

Payment on termination **93**

93.1 The amount due on termination includes (A1)

- an amount due assessed as for normal payments,
- the Defined Cost for Plant and Materials
  - within the Working Areas or
  - to which the *Employer* has title and of which the *Contractor* has to accept delivery,
- other Defined Cost reasonably incurred in expectation of completing the whole of the *works*,
- any amounts retained by the *Employer* and
- a deduction of any un-repaid balance of an advanced payment.

93.2 The amount due on termination also includes one or more of the following as set out in the Termination Table.

A2 The forecast Defined Cost of removing the Equipment.

A3 A deduction of the forecast of the additional cost to the *Employer* of completing the whole of the *works*.

A4 The *direct fee percentage* applied to any excess of the first forecast of the Defined Cost for the *works* over the Price for Work Done to Date less the Fee.

DISPUTE RESOLUTION

Option W1

Dispute resolution procedure (used unless the United Kingdom Housing Grants, Construction and Regeneration Act 1996 applies).

core clauses | main option clauses | secondary option clauses | cost components | contract data

Dispute resolution **W1**

W1.1 A dispute arising under or in connection with this contract is referred to and decided by the *Adjudicator*.

The *Adjudicator* W1.2 (1) The Parties appoint the *Adjudicator* under the NEC Adjudicator's Contract current at the *starting date*.

(2) The *Adjudicator* acts impartially and decides the dispute as an independent adjudicator and not as an arbitrator.

(3) If the *Adjudicator* is not identified in the Contract Data or if the *Adjudicator* resigns or is unable to act, the Parties choose a new adjudicator jointly. If the Parties have not chosen an adjudicator, either Party may ask the *Adjudicator nominating body* to choose one. The *Adjudicator nominating body* chooses an adjudicator within four days of the request. The chosen adjudicator becomes the *Adjudicator*.

(4) A replacement *Adjudicator* has the power to decide a dispute referred to his predecessor but not decided at the time when the predecessor resigned or became unable to act. He deals with an undecided dispute as if it had been referred to him on the date he was appointed.

(5) The *Adjudicator*, his employees and agents are not liable to the Parties for any action or failure to take action in an adjudication unless the action or failure to take action was in bad faith.

The adjudication W1.3 (1) Disputes are notified and referred to the *Adjudicator* in accordance with the Adjudication Table.

ADJUDICATION TABLE

Dispute about	**Which Party may refer it to the *Adjudicator*?**	**When may it be referred to the *Adjudicator*?**
An action of the *Project Manager* or the *Supervisor*	The *Contractor*	Between two and four weeks after the *Contractor*'s notification of the dispute to the *Employer* and the *Project Manager*, the notification itself being made not more than four weeks after the *Contractor* becomes aware of the action
The *Project Manager* or *Supervisor* not having taken an action	The *Contractor*	Between two and four weeks after the *Contractor*'s notification of the dispute to the *Employer* and the *Project Manager*, the notification itself being made not more than four weeks after the *Contractor* becomes aware that the action was not taken
A quotation for a compensation event which is treated as having been accepted	The *Employer*	Between two and four weeks after the *Project Manager*'s notification of the dispute to the *Employer* and the *Contractor,* the notification itself being made not more than four weeks after the quotation was treated as accepted
Any other matter	Either Party	Between two and four weeks after notification of the dispute to the other Party and the *Project Manager*

(2) The times for notifying and referring a dispute may be extended by the *Project Manager* if the *Contractor* and the *Project Manager* agree to the extension before the notice or referral is due. The *Project Manager* notifies the extension that has been agreed to the *Contractor.* If a disputed matter is not notified and referred within the times set out in this contract, neither Party may subsequently refer it to the *Adjudicator* or the *tribunal*.

(3) The Party referring the dispute to the *Adjudicator* includes with his referral information to be considered by the *Adjudicator*. Any more information from a Party to be considered by the *Adjudicator* is provided within four weeks of the referral. This period may be extended if the *Adjudicator* and the Parties agree.

(4) If a matter disputed by the *Contractor* under or in connection with a subcontract is also a matter disputed under or in connection with this contract and if the subcontract allows, the *Contractor* may refer the subcontract dispute to the *Adjudicator* at the same time as the main contract referral. The *Adjudicator* then decides the disputes together and references to the Parties for the purposes of the dispute are interpreted as including the Subcontractor.

(5) The *Adjudicator* may

- review and revise any action or inaction of the *Project Manager* or *Supervisor* related to the dispute and alter a quotation which has been treated as having been accepted,
- take the initiative in ascertaining the facts and the law related to the dispute,
- instruct a Party to provide further information related to the dispute within a stated time and
- instruct a Party to take any other action which he considers necessary to reach his decision and to do so within a stated time.

(6) A communication between a Party and the *Adjudicator* is communicated to the other Party at the same time.

(7) If the *Adjudicator*'s decision includes assessment of additional cost or delay caused to the *Contractor*, he makes his assessment in the same way as a compensation event is assessed.

(8) The *Adjudicator* decides the dispute and notifies the Parties and the *Project Manager* of his decision and his reasons within four weeks of the end of the period for receiving information. This four week period may be extended if the Parties agree.

(9) Unless and until the *Adjudicator* has notified the Parties of his decision, the Parties, the *Project Manager* and the *Supervisor* proceed as if the matter disputed was not disputed.

(10) The *Adjudicator*'s decision is binding on the Parties unless and until revised by the *tribunal* and is enforceable as a matter of contractual obligation between the Parties and not as an arbitral award. The *Adjudicator*'s decision is final and binding if neither Party has notified the other within the times required by this contract that he is dissatisfied with a decision of the *Adjudicator* and intends to refer the matter to the *tribunal*.

(11) The *Adjudicator* may, within two weeks of giving his decision to the Parties, correct any clerical mistake or ambiguity.

Review by the *tribunal* W1.4

(1) A Party does not refer any dispute under or in connection with this contract to the *tribunal* unless it has first been referred to the *Adjudicator* in accordance with this contract.

(2) If, after the *Adjudicator* notifies his decision a Party is dissatisfied, he may notify the other Party that he intends to refer it to the *tribunal*. A Party may not refer a dispute to the *tribunal* unless this notification is given within four weeks of notification of the *Adjudicator*'s decision.

(3) If the *Adjudicator* does not notify his decision within the time provided by this contract, a Party may notify the other Party that he intends to refer the dispute to the *tribunal*. A Party may not refer a dispute to the *tribunal* unless this notification is given within four weeks of the date by which the *Adjudicator* should have notified his decision.

(4) The *tribunal* settles the dispute referred to it. The *tribunal* has the powers to reconsider any decision of the *Adjudicator* and review and revise any action or inaction of the *Project Manager* or the *Supervisor* related to the dispute. A Party is not limited in the *tribunal* proceedings to the information, evidence or arguments put to the *Adjudicator*.

(5) If the *tribunal* is arbitration, the *arbitration procedure*, the place where the arbitration is to be held and the method of choosing the arbitrator are those stated in the Contract Data.

(6) A Party does not call the *Adjudicator* as a witness in *tribunal* proceedings.

Option W2

Dispute resolution procedure (used in the United Kingdom when the Housing Grants, Construction and Regeneration Act 1996 applies).

Dispute resolution **W2**

W2.1 (1) A dispute arising under or in connection with this contract is referred to and decided by the *Adjudicator*. A Party may refer a dispute to the *Adjudicator* at any time.

(2) In this Option, time periods stated in days exclude Christmas Day, Good Friday and bank holidays.

The *Adjudicator* W2.2 (1) The Parties appoint the *Adjudicator* under the NEC Adjudicator's Contract current at the *starting date*.

(2) The *Adjudicator* acts impartially and decides the dispute as an independent adjudicator and not as an arbitrator.

(3) If the *Adjudicator* is not identified in the Contract Data or if the *Adjudicator* resigns or becomes unable to act

- the Parties may choose an adjudicator jointly or
- a Party may ask the *Adjudicator nominating body* to choose an adjudicator.

The *Adjudicator nominating body* chooses an adjudicator within four days of the request. The chosen adjudicator becomes the *Adjudicator.*

(4) A replacement *Adjudicator* has the power to decide a dispute referred to his predecessor but not decided at the time when his predecessor resigned or became unable to act. He deals with an undecided dispute as if it had been referred to him on the date he was appointed.

(5) The *Adjudicator,* his employees and agents are not liable to the Parties for any action or failure to take action in an adjudication unless the action or failure to take action was in bad faith.

The adjudication W2.3 (1) Before a Party refers a dispute to the *Adjudicator*, he gives a notice of adjudication to the other Party with a brief description of the dispute and the decision which he wishes the *Adjudicator* to make. If the *Adjudicator* is named in the Contract Data, the Party sends a copy of the notice of adjudication to the *Adjudicator* when it is issued. Within three days of the receipt of the notice of adjudication, the *Adjudicator* notifies the Parties

- that he is able to decide the dispute in accordance with the contract or
- that he is unable to decide the dispute and has resigned.

If the *Adjudicator* does not so notify within three days of the issue of the notice of adjudication, either Party may act as if he has resigned.

(2) Within seven days of a Party giving a notice of adjudication he

- refers the dispute to the *Adjudicator*,
- provides the *Adjudicator* with the information on which he relies, including any supporting documents and
- provides a copy of the information and supporting documents he has provided to the *Adjudicator* to the other Party.

Any further information from a Party to be considered by the *Adjudicator* is provided within fourteen days of the referral. This period may be extended if the *Adjudicator* and the Parties agree.

(3) If a matter disputed by the *Contractor* under or in connection with a subcontract is also a matter disputed under or in connection with this contract, the *Contractor* may, with the consent of the Subcontractor, refer the subcontract dispute to the *Adjudicator* at the same time as the main contract referral. The *Adjudicator* then decides the disputes together and references to the Parties for the purposes of the dispute are interpreted as including the Subcontractor.

(4) The *Adjudicator* may

- review and revise any action or inaction of the *Project Manager* or *Supervisor* related to the dispute and alter a quotation which has been treated as having been accepted,
- take the initiative in ascertaining the facts and the law related to the dispute,
- instruct a Party to provide further information related to the dispute within a stated time and
- instruct a Party to take any other action which he considers necessary to reach his decision and to do so within a stated time.

(5) If a Party does not comply with any instruction within the time stated by the *Adjudicator*, the *Adjudicator* may continue the adjudication and make his decision based upon the information and evidence he has received.

(6) A communication between a Party and the *Adjudicator* is communicated to the other Party at the same time.

(7) If the *Adjudicator's* decision includes assessment of additional cost or delay caused to the *Contractor,* he makes his assessment in the same way as a compensation event is assessed.

(8) The *Adjudicator* decides the dispute and notifies the Parties and the *Project Manager* of his decision and his reasons within twenty-eight days of the dispute being referred to him. This period may be extended by up to fourteen days with the consent of the referring Party or by any other period agreed by the Parties.

(9) Unless and until the *Adjudicator* has notified the Parties of his decision, the Parties, the *Project Manager* and the *Supervisor* proceed as if the matter disputed was not disputed.

(10) If the *Adjudicator* does not make his decision and notify it to the Parties within the time provided by this contract, the Parties and the *Adjudicator* may agree to extend the period for making his decision. If they do not agree to an extension, either Party may act as if the *Adjudicator* has resigned.

(11) The *Adjudicator*'s decision is binding on the Parties unless and until revised by the *tribunal* and is enforceable as a matter of contractual obligation between the Parties and not as an arbitral award. The *Adjudicator*'s decision is final and binding if neither Party has notified the other within the times required by this contract that he is dissatisfied with a matter decided by the *Adjudicator* and intends to refer the matter to the *tribunal*.

(12) The *Adjudicator* may, within fourteen days of giving his decision to the Parties, correct a clerical mistake or ambiguity.

Review by the *tribunal* W2.4

(1) A Party does not refer any dispute under or in connection with this contract to the *tribunal* unless it has first been decided by the *Adjudicator* in accordance with this contract.

(2) If, after the *Adjudicator* notifies his decision a Party is dissatisfied, that Party may notify the other Party of the matter which he disputes and state that he intends to refer it to the *tribunal*. The dispute may not be referred to the *tribunal* unless this notification is given within four weeks of the notification of the *Adjudicator's* decision.

(3) The *tribunal* settles the dispute referred to it. The *tribunal* has the powers to reconsider any decision of the *Adjudicator* and to review and revise any action or inaction of the *Project Manager* or the *Supervisor* related to the dispute. A Party is not limited in *tribunal* proceedings to the information or evidence put to the *Adjudicator*.

(4) If the *tribunal* is arbitration, the *arbitration procedure,* the place where the arbitration is to be held and the method of choosing the arbitrator are those stated in the Contract Data.

(5) A Party does not call the *Adjudicator* as a witness in *tribunal* proceedings.

SECONDARY OPTION CLAUSES

Option X2: Changes in the law

Changes in the law **X2**

X2.1 A change in the law of the country in which the Site is located is a compensation event if it occurs after the Contract Date. The *Project Manager* may notify the *Contractor* of a compensation event for a change in the law and instruct him to submit quotations. If the effect of a compensation event which is a change in the law is to reduce the total Defined Cost, the Prices are reduced.

Option X4: Parent company guarantee

Parent company guarantee **X4**

X4.1 If a parent company owns the *Contractor*, the *Contractor* gives to the *Employer* a guarantee by the parent company of the *Contractor*'s performance in the form set out in the Works Information. If the guarantee was not given by the Contract Date, it is given to the *Employer* within four weeks of the Contract Date.

Option X5: Sectional Completion

Sectional Completion **X5**

X5.1 In these *conditions of contract*, unless stated as the whole of the *works*, each reference and clause relevant to

- the *works*,
- Completion and
- Completion Date

applies, as the case may be, to either the whole of the *works* or any *section* of the *works*.

Option X6: Bonus for early Completion

Bonus for early Completion **X6**

X6.1 The *Contractor* is paid a bonus calculated at the rate stated in the Contract Data for each day from the earlier of

- Completion and
- the date on which the *Employer* takes over the *works*

until the Completion Date.

Option X7: Delay damages

Delay damages **X7**

X7.1 The *Contractor* pays delay damages at the rate stated in the Contract Data from the Completion Date for each day until the earlier of

- Completion and
- the date on which the *Employer* takes over the *works*.

X7.2 If the Completion Date is changed to a later date after delay damages have been paid, the *Employer* repays the overpayment of damages with interest. Interest is assessed from the date of payment to the date of repayment and the date of repayment is an assessment date.

X7.3 If the *Employer* takes over a part of the *works* before Completion, the delay damages are reduced from the date on which the part is taken over. The *Project Manager* assesses the benefit to the *Employer* of taking over the part of the *works* as a proportion of the benefit to the *Employer* of taking over the whole of the *works* not previously taken over. The delay damages are reduced in this proportion.

Option X12: Partnering

Identified and defined terms **X12**

X12.1 (1) The Partners are those named in the Schedule of Partners. The *Client* is a Partner.

(2) An Own Contract is a contract between two Partners which includes this Option.

(3) The Core Group comprises the Partners listed in the Schedule of Core Group Members.

(4) Partnering Information is information which specifies how the Partners work together and is either in the documents which the Contract Data states it is in or in an instruction given in accordance with this contract.

(5) A Key Performance Indicator is an aspect of performance for which a target is stated in the Schedule of Partners.

Actions X12.2 (1) Each Partner works with the other Partners to achieve the *Client's objective* stated in the Contract Data and the objectives of every other Partner stated in the Schedule of Partners.

(2) Each Partner nominates a representative to act for it in dealings with other Partners.

(3) The Core Group acts and takes decisions on behalf of the Partners on those matters stated in the Partnering Information.

(4) The Partners select the members of the Core Group. The Core Group decides how they will work and decides the dates when each member joins and leaves the Core Group. The *Client's* representative leads the Core Group unless stated otherwise in the Partnering Information.

(5) The Core Group keeps the Schedule of Core Group Members and the Schedule of Partners up to date and issues copies of them to the Partners each time either is revised.

(6) This Option does not create a legal partnership between Partners who are not one of the Parties in this contract.

core clauses | main option clauses | secondary option clauses | cost components | contract data

Working together X12.3 (1) The Partners work together as stated in the Partnering Information and in a spirit of mutual trust and co-operation.

(2) A Partner may ask another Partner to provide information which he needs to carry out the work in his Own Contract and the other Partner provides it.

(3) Each Partner gives an early warning to the other Partners when he becomes aware of any matter that could affect the achievement of another Partner's objectives stated in the Schedule of Partners.

(4) The Partners use common information systems as set out in the Partnering Information.

(5) A Partner implements a decision of the Core Group by issuing instructions in accordance with its Own Contracts.

(6) The Core Group may give an instruction to the Partners to change the Partnering Information. Each such change to the Partnering Information is a compensation event which may lead to reduced Prices.

(7) The Core Group prepares and maintains a timetable showing the proposed timing of the contributions of the Partners. The Core Group issues a copy of the timetable to the Partners each time it is revised. The *Contractor* changes his programme if it is necessary to do so in order to comply with the revised timetable. Each such change is a compensation event which may lead to reduced Prices.

(8) A Partner gives advice, information and opinion to the Core Group and to other Partners when asked to do so by the Core Group. This advice, information and opinion relates to work that another Partner is to carry out under its Own Contract and is given fully, openly and objectively. The Partners show contingency and risk allowances in information about costs, prices and timing for future work.

(9) A Partner notifies the Core Group before subcontracting any work.

Incentives X12.4 (1) A Partner is paid the amount stated in the Schedule of Partners if the target stated for a Key Performance Indicator is improved upon or achieved. Payment of the amount is due when the target has been improved upon or achieved and is made as part of the amount due in the Partner's Own Contract.

(2) The *Client* may add a Key Performance Indicator and associated payment to the Schedule of Partners but may not delete or reduce a payment stated in the Schedule of Partners.

Option X13: Performance bond

Performance bond **X13**

X13.1 The *Contractor* gives the *Employer* a performance bond, provided by a bank or insurer which the *Project Manager* has accepted, for the amount stated in the Contract Data and in the form set out in the Works Information. A reason for not accepting the bank or insurer is that its commercial position is not strong enough to carry the bond. If the bond was not given by the Contract Date, it is given to the *Employer* within four weeks of the Contract Date.

Option X14: Advanced payment to the *Contractor*

Advanced payment **X14**

X14.1 The *Employer* makes an advanced payment to the *Contractor* of the amount stated in the Contract Data.

X14.2 The advanced payment is made either within four weeks of the Contract Date or, if an advanced payment bond is required, within four weeks of the later of

- the Contract Date and
- the date when the *Employer* receives the advanced payment bond.

The advanced payment bond is issued by a bank or insurer which the *Project Manager* has accepted. A reason for not accepting the proposed bank or insurer is that its commercial position is not strong enough to carry the bond. The bond is for the amount of the advanced payment which the *Contractor* has not repaid and is in the form set out in the Works Information. Delay in making the advanced payment is a compensation event.

X14.3 The advanced payment is repaid to the *Employer* by the *Contractor* in instalments of the amount stated in the Contract Data. An instalment is included in each amount due assessed after the period stated in the Contract Data has passed until the advanced payment has been repaid.

Option X15: Limitation of the *Contractor*'s liability for his design to reasonable skill and care

The *Contractor*'s design **X15**

X15.1 The *Contractor* is not liable for Defects in the *works* due to his design so far as he proves that he used reasonable skill and care to ensure that his design complied with the Works Information.

X15.2 If the *Contractor* corrects a Defect for which he is not liable under this contract it is a compensation event.

Option X16: Retention

Retention **X16**

X16.1 After the Price for Work Done to Date has reached the *retention free amount*, an amount is retained in each amount due. Until the earlier of

- Completion of the whole of the *works* and
- the date on which the *Employer* takes over the whole of the *works*

the amount retained is the *retention percentage* applied to the excess of the Price for Work Done to Date above the *retention free amount*.

X16.2 The amount retained is halved

- in the assessment made at Completion of the whole of the *works* or
- in the next assessment after the *Employer* has taken over the whole of the *works* if this is before Completion of the whole of the *works*.

The amount retained remains at this amount until the Defects Certificate is issued. No amount is retained in the assessments made after the Defects Certificate has been issued.

Option X17: Low performance damages

Low performance damages **X17**

X17.1 If a Defect included in the Defects Certificate shows low performance with respect to a performance level stated in the Contract Data, the *Contractor* pays the amount of low performance damages stated in the Contract Data.

Option X18: Limitation of liability

Limitation of liability **X18**

X18.1 The *Contractor*'s liability to the *Employer* for the *Employer*'s indirect or consequential loss is limited to the amount stated in the Contract Data.

X18.2 For any one event, the liability of the *Contractor* to the *Employer* for loss of or damage to the *Employer*'s property is limited to the amount stated in the Contract Data.

X18.3 The *Contractor*'s liability to the *Employer* for Defects due to his design which are not listed on the Defects Certificate is limited to the amount stated in the Contract Data.

X18.4 The *Contractor*'s total liability to the *Employer* for all matters arising under or in connection with this contract, other than the excluded matters, is limited to the amount stated in the Contract Data and applies in contract, tort or delict and otherwise to the extent allowed under the *law of the contract.*

The excluded matters are amounts payable by the *Contractor* as stated in this contract for

- loss of or damage to the *Employer*'s property,
- delay damages if Option X7 applies and
- low performance damages if Option X17 applies.

X18.5 The *Contractor* is not liable to the *Employer* for a matter unless it is notified to the *Contractor* before the *end of liability date.*

Option X20: Key Performance Indicators (not used with Option X12)

Incentives

X20.1 A Key Performance Indicator is an aspect of performance by the *Contractor* for which a target is stated in the Incentive Schedule. The Incentive Schedule is the *incentive schedule* unless later changed in accordance with this contract.

X20.2 From the *starting date* until the Defects Certificate has been issued, the *Contractor* reports to the *Project Manager* his performance against each of the Key Performance Indicators. Reports are provided at the intervals stated in the Contract Data and include the forecast final measurement against each indicator.

X20.3 If the *Contractor's* forecast final measurement against a Key Performance Indicator will not achieve the target stated in the Incentive Schedule, he submits to the *Project Manager* his proposals for improving performance.

X20.4 The *Contractor* is paid the amount stated in the Incentive Schedule if the target stated for a Key Performance Indicator is improved upon or achieved. Payment of the amount is due when the target has been improved upon or achieved.

X20.5 The *Employer* may add a Key Performance Indicator and associated payment to the Incentive Schedule but may not delete or reduce a payment stated in the Incentive Schedule.

OPTION Y

Option Y(UK)2: The Housing Grants, Construction and Regeneration Act 1996

Definitions **Y(UK)2**

Y2.1 (1) The Act is The Housing Grants, Construction and Regeneration Act 1996.

(2) A period of time stated in days is a period calculated in accordance with Section 116 of the Act.

Dates for payment Y2.2 The date on which a payment becomes due is seven days after the assessment date.

The final date for payment is fourteen days or a different period for payment if stated in the Contract Data after the date on which payment becomes due.

The *Project Manager*'s certificate is the notice of payment from the *Employer* to the *Contractor* specifying the amount of the payment made or proposed to be made and stating how the amount was calculated.

Notice of intention to withhold payment Y2.3 If either Party intends to withhold payment of an amount due under this contract, he notifies the other Party not later than seven days (the prescribed period) before the final date for payment by stating the amount proposed to be withheld and the reason for withholding payment. If there is more than one reason, the amount for each reason is stated.

A Party does not withhold payment of an amount due under this contract unless he has notified his intention to withhold payment as required by this contract.

Suspension of performance Y2.4 If the *Contractor* exercises his right under the Act to suspend performance, it is a compensation event.

Option Y(UK)3: The Contracts (Rights of Third Parties) Act 1999

Third party rights **Y(UK)3**

Y3.1 A person or organisation who is not one of the Parties may enforce a term of this contract under the Contracts (Rights of Third Parties) Act 1999 only if the term and the person or organisation are stated in the Contract Data.

Option Z: *Additional conditions of contract*

Additional conditions of contract **Z1**

Z1.1 The *additional conditions of contract* stated in the Contract Data are part of this contract.

core clauses | main option clauses | secondary option clauses | cost components | contract data

SCHEDULE OF COST COMPONENTS

In this schedule the *Contractor* means the *Contractor* and not his Subcontractors. An amount is included only in one cost component and only if it is incurred in order to Provide the Works.

People 1 The following components of the cost of

- people who are directly employed by the *Contractor* and whose normal place of working is within the Working Areas and
- people who are directly employed by the *Contractor* and whose normal place of working is not within the Working Areas but who are working in the Working Areas.

11 Wages, salaries and amounts paid by the *Contractor* for people paid according to the time worked while they are within the Working Areas.

12 Payments to people for

(a) bonuses and incentives
(b) overtime
(c) working in special circumstances
(d) special allowances
(e) absence due to sickness and holidays
(f) severance related to work on this contract.

13 Payments made in relation to people for

(a) travel
(b) subsistence and lodging
(c) relocation
(d) medical examinations
(e) passports and visas
(f) travel insurance
(g) items (a) to (f) for dependants
(h) protective clothing
(i) meeting the requirements of the law
(j) pensions and life assurance
(k) death benefit
(l) occupational accident benefits
(m) medical aid
(n) a vehicle
(o) safety training.

14 The following components of the cost of people who are not directly employed by the *Contractor* but are paid for by him according to the time worked while they are within the Working Areas.

Amounts paid by the *Contractor*.

Equipment 2 The following components of the cost of Equipment which is used within the Working Areas (including the cost of accommodation but excluding Equipment cost covered by the percentage for Working Areas overheads).

21 Payments for the hire or rent of Equipment not owned by

- the *Contractor*,
- his parent company or
- by a company with the same parent company

at the hire or rental rate multiplied by the time for which the Equipment is required.

22 Payments for Equipment which is not listed in the Contract Data but is

- owned by the *Contractor*,
- purchased by the *Contractor* under a hire purchase or lease agreement or
- hired by the *Contractor* from the *Contractor*'s parent company or from a company with the same parent company

at open market rates, multiplied by the time for which the Equipment is required.

23 Payments for Equipment purchased for work included in this contract listed with a time-related on cost charge, in the Contract Data, of

- the change in value over the period for which the Equipment is required and
- the time-related on cost charge stated in the Contract Data for the period for which the Equipment is required.

The change in value is the difference between the purchase price and either the sale price or the open market sale price at the end of the period for which the Equipment is required. Interim payments of the change in value are made at each assessment date. A final payment is made in the next assessment after the change in value has been determined.

If the *Project Manager* agrees, an additional item of Equipment may be assessed as if it had been listed in the Contract Data.

24 Payments for special Equipment listed in the Contract Data. These amounts are the rates stated in the Contract Data multiplied by the time for which the Equipment is required.

If the *Project Manager* agrees, an additional item of special Equipment may be assessed as if it had been listed in the Contract Data.

25 Payments for the purchase price of Equipment which is consumed.

26 Unless included in the hire or rental rates, payments for

- transporting Equipment to and from the Working Areas other than for repair and maintenance,
- erecting and dismantling Equipment and
- constructing, fabricating or modifying Equipment as a result of a compensation event.

27 Payments for purchase of materials used to construct or fabricate Equipment.

28 Unless included in the hire rates, the cost of operatives is included in the cost of people.

Plant and Materials 3 The following components of the cost of Plant and Materials.

31 Payments for

- purchasing Plant and Materials,
- delivery to and removal from the Working Areas,
- providing and removing packaging and
- samples and tests.

32 Cost is credited with payments received for disposal of Plant and Materials unless the cost is disallowed.

Charges 4 The following components of the cost of charges paid by the *Contractor*.

41 Payments for provision and use in the Working Areas of

- water,
- gas and
- electricity.

42 Payments to public authorities and other properly constituted authorities of charges which they are authorised to make in respect of the *works*.

43 Payments for

- (a) cancellation charges arising from a compensation event
- (b) buying or leasing land
- (c) compensation for loss of crops or buildings
- (d) royalties
- (e) inspection certificates
- (f) charges for access to the Working Areas
- (g) facilities for visits to the Working Areas by Others
- (h) specialist services
- (i) consumables and equipment provided by the *Contractor* for the *Project Manager*'s and *Supervisor*'s offices.

44 A charge for overhead costs incurred within the Working Areas calculated by applying the percentage for Working Areas overheads stated in the Contract Data to the total of people items 11, 12, 13 and 14. The charge includes provision and use of equipment, supplies and services, but excludes accommodation, for

- (a) catering
- (b) medical facilities and first aid
- (c) recreation
- (d) sanitation
- (e) security
- (f) copying
- (g) telephone, telex, fax, radio and CCTV
- (h) surveying and setting out
- (i) computing
- (j) hand tools not powered by compressed air.

Manufacture and fabrication

5 The following components of the cost of manufacture and fabrication of Plant and Materials which are

- wholly or partly designed specifically for the *works* and
- manufactured or fabricated outside the Working Areas.

51 The total of the hours worked by employees multiplied by the hourly rates stated in the Contract Data for the categories of employees listed.

52 An amount for overheads calculated by multiplying this total by the percentage for manufacturing and fabrication overheads stated in the Contract Data.

Design

6 The following components of the cost of design of the *works* and Equipment done outside the Working Areas.

61 The total of the hours worked by employees multiplied by the hourly rates stated in the Contract Data for the categories of employees listed.

62 An amount for overheads calculated by multiplying this total by the percentage for design overheads stated in the Contract Data.

63 The cost of travel to and from the Working Areas for the categories of design employees listed in the Contract Data.

Insurance

7 The following are deducted from cost

- the cost of events for which this contract requires the *Contractor* to insure and
- other costs paid to the *Contractor* by insurers.

SHORTER SCHEDULE OF COST COMPONENTS

An amount is included only in one cost component and only if it is incurred in order to Provide the Works.

People **1** The following components of the cost of

- people who are directly employed by the *Contractor* and whose normal place of working is within the Working Areas,
- people who are directly employed by the *Contractor* and whose normal place of working is not within the Working Areas but who are working in the Working Areas and
- people who are not directly employed by the *Contractor* but are paid for by him according to the time worked while they are within the Working Areas.

11 Amounts paid by the *Contractor* including those for meeting the requirements of the law and for pension provision.

Equipment **2** The following components of the cost of Equipment which is used within the Working Areas (including the cost of accommodation but excluding Equipment cost covered by the percentage for people overheads).

21 Amounts for Equipment which is in the published list stated in the Contract Data. These amounts are calculated by applying the percentage adjustment for listed Equipment stated in the Contract Data to the rates in the published list and by multiplying the resulting rate by the time for which the Equipment is required.

22 Amounts for Equipment listed in the Contract Data which is not in the published list stated in the Contract Data. These amounts are the rates stated in the Contract Data multiplied by the time for which the Equipment is required.

23 The time required is expressed in hours, days, weeks or months consistently with the list of items of Equipment in the Contract Data or with the published list stated in the Contract Data.

24 Unless the item is in the published list and the rate includes the cost component, payments for

- transporting Equipment to and from the Working Areas other than for repair and maintenance,
- erecting and dismantling Equipment and
- constructing, fabricating or modifying Equipment as a result of a compensation event.

25 Unless the item is in the published list and the rate includes the cost component, the purchase price of Equipment which is consumed.

26 Unless included in the rate in the published list, the cost of operatives is included in the cost of people.

27 Amounts for Equipment which is neither in the published list stated in the Contract Data nor listed in the Contract Data, at competitively tendered or open market rates, multiplied by the time for which the Equipment is required.

Plant and Materials **3** The following components of the cost of Plant and Materials.

31 Payments for

- purchasing Plant and Materials,
- delivery to and removal from the Working Areas,
- providing and removing packaging and
- samples and tests.

32 Cost is credited with payments received for disposal of Plant and Materials unless the cost is disallowed.

Charges **4** The following components of the cost of charges paid by the *Contractor.*

41 A charge calculated by applying the percentage for people overheads stated in the Contract Data to people item 11 to cover the costs of

- payments for the provision and use in the Working Areas of water, gas and electricity,
- payments for buying or leasing land, compensation for loss of crops or buildings, royalties, inspection certificates, charges for access to the Working Areas, facilities for visits to the Working Areas by Others and
- payments for equipment, supplies and services for offices, drawing office, laboratories, workshops, stores and compounds, labour camps, cabins, catering, medical facilities and first aid, recreation, sanitation, security, copying, telephone, telex, fax, radio, CCTV, surveying and setting out, computing, and hand tools not powered by compressed air.

42 Payments for cancellation charges arising from a compensation event.

43 Payments to public authorities and other properly constituted authorities of charges which they are authorised to make in respect of the *works.*

44 Consumables and equipment provided by the *Contractor* for the *Project Manager*'s and *Supervisor*'s office.

45 Specialist services.

Manufacture and fabrication **5** The following components of the cost of manufacture and fabrication of Plant and Materials, which are

- wholly or partly designed specifically for the *works* and
- manufactured or fabricated outside the Working Areas.

51 Amounts paid by the *Contractor.*

Design **6** The following components of the cost of design of the *works* and Equipment done outside the Working Areas.

61 The total of the hours worked by employees multiplied by the hourly rates stated in the Contract Data for the categories of employees listed.

62 An amount for overheads calculated by multiplying this total by the percentage for design overheads stated in the Contract Data.

63 The cost of travel to and from the Working Areas for the categories of design employees listed in the Contract Data.

Insurance **7** The following are deducted from cost

- costs against which this contract required the *Contractor* to insure and
- other costs paid to the *Contractor* by insurers.

CONTRACT DATA

Part one – Data provided by the *Employer*

Completion of the data in full, according to the Options chosen, is essential to create a complete contract.

Statements given in all contracts

1 General

- The *conditions of contract* are the core clauses and the clauses for main Option E, dispute resolution Option ..... and secondary Options ..... of the NEC3 Engineering and Construction Contract June 2005 (with amendments June 2006).
- The *works* are

  ....

- The *Employer* is

  Name ....

  Address ....

  ....

- The *Project Manager* is

  Name ....

  Address ....

  ....

- The *Supervisor* is

  Name ....

  Address ....

  ....

- The *Adjudicator* is

  Name ....

  Address ....

  ....

- The Works Information is in

  ....

  ....

  ....

- The Site Information is in

  ....

  ....

  ....

core clauses | main option clauses | secondary option clauses | cost components | contract data

- The *boundaries of the site* are. . . . .
- The *language of this contract* is . . . .
- The *law of the contract* is the law of . . . .
- The *period for reply* is . . . . weeks.
- The *Adjudicator nominating body* is . . . .
- The *tribunal* is . . . .
  . . . .
- The following matters will be included in the Risk Register
  . . . .
  . . . .
  . . . .

3 Time

- The *starting date* is . . . .
- The *access dates* are

Part of the Site	Date
1	. . . .
2	. . . .
3	. . . .

- The *Contractor* submits revised programmes at intervals no longer than . . . . weeks.

4 Testing and Defects

- The *defects date* is . . . . . weeks after Completion of the whole of the *works*.
- The *defect correction period* is . . . . weeks except that
  - The *defect correction period* for . . . . is . . . . weeks
  - The *defect correction period* for . . . . is . . . . weeks.

5 Payment

- The *currency of this contract* is the. . . . .
- The *assessment interval* is . . . . weeks (not more than five).
- The *interest rate* is . . . . . % per annum (not less than 2) above the . . . . . . rate of the . . . . bank.

6 Compensation events

- The place where weather is to be recorded is

  ....

- The *weather measurements* to be recorded for each calendar month are
  - the cumulative rainfall (mm)
  - the number of days with rainfall more than 5 mm
  - the number of days with minimum air temperature less than 0 degrees Celsius
  - the number of days with snow lying at .... hours GMT
  - and these measurements:

    ....

    ....

    ....

- The *weather measurements* are supplied by ....
- The *weather data* are the records of past *weather measurements* for each calendar month which were recorded at ....

  and which are available from ....

  ....

Where no recorded data are available

- Assumed values for the ten year return *weather data* for each *weather measurement* for each calendar month are

  ....

  ....

  ....

  ....

8 Risks and insurance

- The minimum limit of indemnity for insurance in respect of loss of or damage to property (except the *works*, Plant and Materials and Equipment) and liability for bodily injury to or death of a person (not an employee of the *Contractor*) caused by activity in connection with this contract for any one event is

  ....

- The minimum limit of indemnity for insurance in respect of death of or bodily injury to employees of the *Contractor* arising out of and in the course of their employment in connection with this contract for any one event is

  ....

Optional statements

If the *tribunal* is arbitration

- The *arbitration procedure* is ....
- The place where arbitration is to be held is

  ....

- The person or organisation who will choose an arbitrator
  - if the Parties cannot agree a choice or
  - if the *arbitration procedure* does not state who selects an arbitrator is

  ....

If the *Employer* has decided the *completion date* for the whole of the *works*

- The *completion date* for the whole of the *works* is ....

If the *Employer* is not willing to take over the *works* before the Completion Date

- The *Employer* is not willing to take over the *works* before the Completion Date.

If no programme is identified in part two of the Contract Data

- The *Contractor* is to submit a first programme for acceptance within .... weeks of the Contract Date.

If the *Employer* has identified work which is to meet a stated *condition* by a *key date*

- The *key dates* and *conditions* to be met are

condition to be met	*key date*
1	....
2	....
3	....

If the period in which payments are made is not three weeks and Y(UK)2 is not used

- The period within which payments are made is ....

If Y(UK)2 is used and the final date for payment is not 14 days after the date when payment is due

- The period for payment is ....

If there are additional *Employer*'s risks

- These are additional *Employer*'s risks

  1 ....

  2 ....

  3 ....

If the *Employer* is to provide Plant and Materials

- The insurance against loss of or damage to the *works*, Plant and Materials is to include cover for Plant and Materials provided by the *Employer* for an amount of

  ....

If the *Employer* is to provide any of the insurances stated in the Insurance Table

- The *Employer* provides these insurances from the Insurance Table

  1. Insurance against. . . . .

  Cover/indemnity is . . . .

  The deductibles are. . . . .

  2. Insurance against. . . . .

  Cover/indemnity is . . . .

  The deductibles are. . . . .

  3. Insurance against. . . . .

  Cover/indemnity is . . . .

  The deductibles are. . . . .

If additional insurances are to be provided

- The *Employer* provides these additional insurances

  1. Insurance against. . . . .

  Cover/indemnity is . . . .

  The deductibles are. . . . .

  2. Insurance against. . . . .

  Cover/indemnity is . . . .

  The deductibles are. . . . .

  3. Insurance against. . . . .

  Cover/indemnity is . . . .

  The deductibles are. . . . .

- The *Contractor* provides these additional insurances

  1. Insurance against. . . . .

  Cover/indemnity is . . . .

  2. Insurance against. . . . .

  Cover/indemnity is . . . .

  3. Insurance against. . . . .

  Cover/indemnity is . . . .

- **The *Contractor* prepares forecasts of Defined Cost for the *works* at intervals no longer than. . . . . weeks.**
- **The *exchange rates* are those published in. . . . . on . . . . (date).**

If Option X5 is used

- The *completion date* for each *section* of the *works* is

section	description	*completion date*
1	....	....
2	....	....
3	....	....
4	....	....

If Options X5 and X6 are used together

- The bonus for each *section* of the *works* is

section	description	amount per day
1	....	....
2	....	....
3	....	....
4	....	....

Remainder of the *works* ....

If Options X5 and X7 are used together

- Delay damages for each *section* of the *works* are

section	description	amount per day
1	....	....
2	....	....
3	....	....
4	....	....

Remainder of the *works* ....

If Option X6 is used (but not if Option X5 is also used)

- The bonus for the whole of the *works* is .... per day.

If Option X7 is used (but not if Option X5 is also used)

- Delay damages for Completion of the whole of the *works* are ...... per day.

If Option X12 is used

- The *Client* is

  Name ....

  Address ....

  ....

- The *Client's objective* is

  ....

  ....

  ....

  ....

  ....

  ....

- The Partnering Information is in

  ....

  ....

  ....

  ....

If Option X13 is used

- The amount of the performance bond is ....

If Option X14 is used

- The amount of the advanced payment is ....
- The *Contractor* repays the instalments in assessments starting not less than .... weeks after the Contract Date.
- The instalments are..

  ....

  (either an amount or a percentage of the payment otherwise due)
- An advanced payment bond is/is not required.

If Option X16 is used

- The *retention free amount* is...
- The *retention percentage* is ....%.

If Option X17 is used

- The amounts for low performance damages are

amount	performance level
....	for ....
....	for ....
....	for ....
....	for ....

If Option X18 is used

- The *Contractor*'s liability to the *Employer* for indirect or consequential loss is limited to ....
- For any one event, the *Contractor*'s liability to the *Employer* for loss of or damage to the *Employer*'s property is limited to.......................
- The *Contractor*'s liability for Defects due to his design which are not listed on the Defects Certificate is limited to ....
- The *Contractor*'s total liability to the *Employer* for all matters arising under or in connection with this contract, other than excluded matters, is limited to..
- The *end of liability date* is............years after the Completion of the whole of the *works*.

If Option X20 is used (but not if Option X12 is also used)

- The *incentive schedule* for Key Performance Indicators is in ....
- A report of performance against each Key Performance Indicator is provided at intervals of .... months.

core clauses | main option clauses | secondary option clauses | cost components | contract data

If Option Y(UK)3 is used

- term — person or organisation

term	person or organisation
....	....
....	....
....	....
....	....

If Option Z is used

- The *additional conditions of contract* are ....
....

Part two – Data provided by the *Contractor*

Completion of the data in full, according to the Options chosen, is essential to create a complete contract.

Statements given in all contracts

- The *Contractor* is

  Name ....

  Address ....

  ....

- The *direct fee percentage* is ....%.
- The *subcontracted fee percentage* is ....%.
- The *working areas* are the Site and ....
- The key people are

  (1) Name...

  Job ....

  Responsibilities...

  ....

  Qualifications ....

  Experience...

  ....

  (2) Name...

  Job ....

  Responsibilities...

  ....

  Qualifications...

  Experience...

  ....

- The following matters will be included in the Risk Register

  ....

  ....

  ....

  ....

Optional statements

If the *Contractor* is to provide Works Information for his design

- The Works Information for the *Contractor*'s design is in

  ....

  ....

  ....

  ....

  ....

  ....

If a programme is to be identified in the Contract Data

- The programme identified in the Contract Data is . . . .

If the *Contractor* is to decide the *completion date* for the whole of the *works*

- The *completion date* for the whole of the *works* is . . . .

Data for Schedule of Cost Components

- **The listed items of Equipment purchased for work on this contract, with an on cost charge, are**

Equipment	time-related charge	per time period
. . . .	. . . .	per . . . .
. . . .	. . . .	per . . . .
. . . .	. . . .	per . . . .
. . . .	. . . .	per . . . .

- **The rates for special Equipment are**

Equipment	size or capacity	rate
. . . .	. . . .	. . . .
. . . .	. . . .	. . . .
. . . .	. . . .	. . . .
. . . .	. . . .	. . . .

- **The percentage for Working Areas overheads is . . . .%.**
- **The hourly rates for Defined Cost of manufacture and fabrication outside the Working Areas are**

category of employee	hourly rate
. . . .	. . . .
. . . .	. . . .
. . . .	. . . .
. . . .	. . . .

- **The percentage for manufacture and fabrication overheads is . . . .%.**

Data for both schedules of cost components

- **The hourly rates for Defined Cost of design outside the Working Areas are**

category of employee	hourly rate
. . . .	. . . .
. . . .	. . . .
. . . .	. . . .
. . . .	. . . .

- **The percentage for design overheads is . . . .%.**

- The categories of design employees whose travelling expenses to and from the Working Areas are included as a cost of design of the *works* and Equipment done outside of the Working Areas are

  ....

  ....

  ....

  ....

Data for the Shorter Schedule of Cost Components

- The percentage for people overheads is ....%.
- The published list of Equipment is the last edition of the list published by

  ....
- The percentage for adjustment for Equipment in the published list is ....% (state plus or minus).
- The rates for other Equipment are

Equipment	size or capacity	rate
....	....	....
....	....	....
....	....	....
....	....	....

nec 3 Engineering and Construction Contract Option E

Index by clause numbers (Option clauses indicated by their letters, main clause heads by bold numbers). Terms in *italics* are identified in Contract Data, and defined terms have capital initial letters.

